Saving Salila's Turtle

An Environmental Engineering Story

Written by the Engineering is Elementary Team

Illustrated by Jeannette Martin

Chapter One | A Rainy Day Discovery

Rainclouds hung dark overhead as Father and I walked home along the banks of the *Ganga Ma*—Mother Ganges. The river currents swirled, making choppy waves in the water.

Turning to the river, I asked, "Are you upset today?"

"Salila," said Father, "you talk to the river so often that I think some day the water might actually answer you. Your mother and I named you well."

I smiled. My name, Salila, is the Hindi word for water. I've always felt a special connection to the river—and to all the plants and animals living in my city.

I looked away from the Ganges as a fat raindrop plopped onto the book I was carrying. It was the beginning of monsoon season. The raindrops fell faster, making dark splotches on my school uniform.

"Come on, Salila, let's hurry," Father said. I turned back to take one last look at the river.

In that moment, something caught my eye. "Wait, Father, look!" I cried as I pointed.

"What is it?" Father asked, scanning the bank of the river.

I reached down and picked up a small *kachua*, a turtle, that was climbing out of the water. I placed it in my hand, sheltering the animal from the rain, which was coming down even harder now. Through my wet bangs, I looked towards the spot where the turtle had emerged. The water was covered with oil that created slick, shimmering rainbows.

"Salila, we have to go," Father said as raindrops trickled off his nose. "Put the *kachua* back in the water."

"I can't leave the turtle here!" I cried. "The water is dirty! Won't it make the turtle sick?"

"Salila, this is the turtle's home. And it's pouring. Maybe you can come back later," said Father. "Let's go."

Reluctantly I left the turtle where I'd found it, making sure I remembered the exact spot so I could return to look for it later. How had I never noticed that my beautiful Ganges, important to people and so many animals, was dirty? As the rain poured down around me, all I could think about was coming back to make sure the turtle was safe.

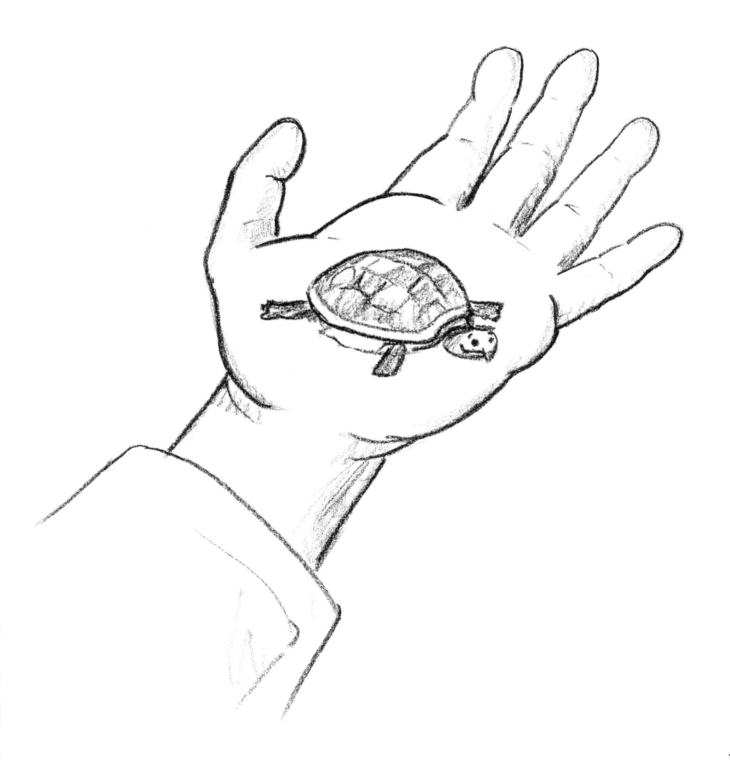

Chapter Two | Pollution Problems

"You're soaked!" my mother cried, standing in the doorway. "Don't track that mud through the house."

"Mother, it was horrible," I said as she started to rub my wet hair with a towel. "On the way home from school we found a *kachua* in the Ganges. The water was oily and dirty. I know it's going to make the turtle sick. I have to rescue the turtle. And what about the other animals living in the water? I need to make the Ganges safe for them!"

"Okay, Salila," Mother said. "Take a deep breath. We'll figure out a way to help your turtle and the river."

I knew my mother would be able to help. She knows a lot about the water in our country, because she's an environmental engineer. She uses her knowledge of science and math and her creativity to help make the air, water, and soil in the environment cleaner and safer.

"Problems like these are exactly why I became an environmental engineer," said Mother. "India is a country surrounded by water, and we have many rivers—yet there are still people who don't have access to clean water. We think of the Ganges as sacred because the water allows us to live. But we need to keep the Ganges clean in order for her to continue giving life—to people and to animals."

"Father," I said, "I knew we should have rescued the *kachua*!"

"It is important to help the environment and the animals who live in it," Father said. "But what would we have done with the turtle?"

"I could keep the turtle in a tank. A big tank, so it would have plenty of room to crawl and swim. Oh, I know!" I said, clapping my hands together. "I can clean some water from the Ganges and use it to fill the tank! Then the turtle will feel at home. I just need to remove the trash from the water."

Mother and Father looked at each other in silence for a few moments. I knew they were thinking about whether I would get to bring home the turtle.

"That sounds like the beginning of a good plan," Mother said, finally.

"I agree," Father nodded. "If you do careful work to make a cleaner habitat for the turtle, then you can keep it as a pet."

I could feel butterflies fluttering in my stomach. My own pet! I knew I'd have to do a good job of cleaning water for the turtle.

"It's important for you to learn about removing contaminants from the Ganges," Mother added, "for the turtle and for the people of India. Tomorrow after school I'll take you to the university to show you the work some of my colleagues have done to clean water."

"Tomorrow?" I asked. "What should I do to get ready?"

"Well," Mother began, "when something is polluted it means it has become filled with harmful substances that dirty the environment. You could start by asking some good questions about how the water in the Ganges becomes polluted. Where does the water in the Ganges come from? How do the pollutants get into the water?"

Father nodded. "Come, Salila," he said. "Together I bet we can answer some of those questions."

Chapter Three | The Water Cycle

As Mother disappeared into her office, I walked to the sink to get a drink. "There's the answer to the first question Mother asked," I said. "Water comes from the faucet. And at some point before it comes from the faucet, it comes from the river."

"Right," Father said. "But what about before that? Do you remember learning about the water cycle in school?"

I took a sip of my water, imagining the travels it had taken to get into my glass. "I remember that we learned about water falling from the sky as rain, and water melting from glaciers in the mountains and flowing into streams. Then some of the water from streams and oceans dries up and goes back into the air," I said. "That's called evaporation."

"Right," Father said. "Now, what about the pollution you saw in the river today? How did the river become polluted?"

"People use the Ganges for lots of things," I said. "I see boats traveling on the river, people cleaning their clothes in the water, and even people taking baths in the river with soap. All of those things must add pollutants."

"Yes," Father said. "Oil from boats and soap from washing are contaminants—things found in the environment that make it unclean or unsafe. People and machines can add contaminants to the water in the river that lead to pollution. But what about when the water isn't in the river? What about when it is moving through the rest of the water cycle?"

"Well . . ." I thought for a few moments. "If water goes through the air and the ground and then into our streams, I bet water can be affected by pollution in the land and air, too. There's that factory across the river that sends smoke into

the air, and trash that gets dumped on the land. Both of those things could pollute the water."

"That's true," Father said. "In order to make our water cleaner, we really need to think of ways to stop all types of pollution."

"When I saw the turtle, I just wanted to rescue it from the water," I said. "I wasn't thinking about cleaning up the whole environment! How can I do that?" I could feel worry rising up from my stomach.

"Don't worry, Salila," said Father. "I know it might seem like an impossible job right now, but even the smallest steps toward a cleaner environment are important. First you can focus on cleaning water for the turtle's tank. Then you can use what you learn to think about ways to clean the water in the Ganges. After you go to Mother's office tomorrow, I bet you'll have some new ideas."

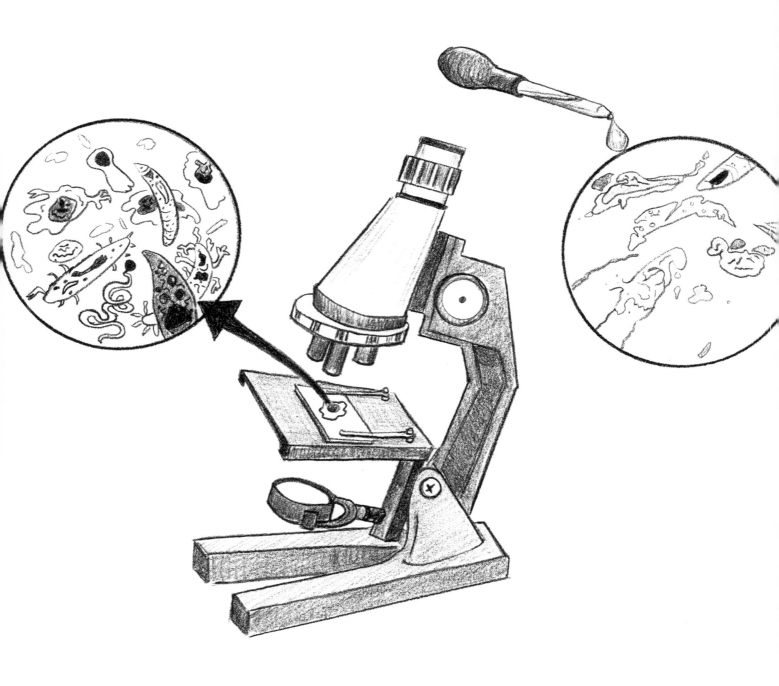

Chapter Four | A Visit to the Lab

The next day after school, I found myself watching strange translucent shapes, wriggling and drifting in a drop of water, through a circle of light.

"The water is alive!" I cried, amazed at the tiny creatures I saw through the microscope.

The man beside me, Dr. Gadgil, chuckled. "Though we can't see them with our eyes, there are lots of things that live in the river water with your turtle. It can be very surprising to see microbes for the first time."

I turned to Mother, who was sitting next to me and nodding. "I wanted you to see the microbes for yourself, Salila," she said. "Those tiny creatures in the water are part of the challenge of keeping water safe and clean."

"That's right," Dr. Gadgil said. "It is just as important to remove the harmful microbes from polluted water as it is to remove the trash that is easy to see. Some of the microbes are tiny plants and animals. Some are another type of living thing called bacteria. Not all of them are harmful, but even one harmful microbe could make you sick."

"How can we get these microbes out of the water when we can't even see them without a microscope?" I asked.

"Well, there are a few ways to do it," said Dr. Gadgil. "In many cities, people have their water cleaned at a purification plant before it comes through the pipes to their homes. Contaminants like dirt, twigs, and microbes are removed with filters, and then a chemical called chlorine is added to kill any of the microbes that are left."

"That's why it's okay for us to drink the water that comes out of our faucets," I said, thinking out loud. "Even though the water comes from a polluted river, it gets cleaned before it comes to our homes. But what about people who don't have filtered water brought into their houses with pipes, like my grandmother? In her village you have to go to the river to get your water."

"That's why Grandmother boils her water," Mother said. "Do you remember helping to build the fire when we visited?" she asked.

I nodded. "But I also remember Grandmother said

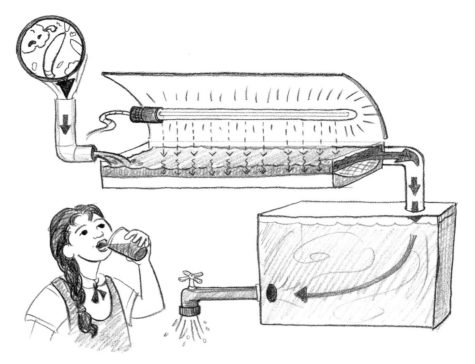

sometimes people drink water without boiling it—if there isn't enough wood to burn."

Dr. Gadgil nodded. "Many villages face that problem. That's why I invented another way to clean water. I call it the Ultraviolet Waterworks. It has a special lightbulb—an ultraviolet lightbulb. The rays of ultraviolet light are like intense sunshine. The light burns and destroys the microbes so they can't harm people."

"So the Ultraviolet Waterworks helps you clean even the things in the water you can't see," I said. "I don't know if I can come up with something that fancy."

"Technology doesn't always have to be fancy," Dr. Gadgil said. "Technology is any thing or process that people create to solve problems. For the water in your turtle's tank, you might be able to use a filter you make at home—something that can separate the trash and impurities from the water. I'm sure you'll come up with a solution that will work well for your turtle."

....

Later that day, as I thought about what Dr. Gadgil and Mother had said, I felt less and less sure that I would be able to clean water from the Ganges for the *kachua*.

I heard Mother chuckling and turned toward her. "I know that look on your face," she said. "You're frustrated."

"How can you tell?" I asked.

"I make that face when I'm frustrated, too," she said. "Now, why are you upset?"

"I still feel stuck," I said to Mother. "I want to help the turtle, and I think filtering water so I can keep the turtle in a tank is a good idea. But I also want to help the whole Ganges. I don't know where to start. Usually when I have a problem, I can come up with a solution right away. I guess I feel like I'm letting down the animals in the Ganges."

"But you're not, Salila," Mother said. "Thinking about helping one animal, and about the problems in our environment, are excellent first steps. Pollution is a big problem. There isn't one easy way to fix it."

"I guess so," I said. "But I'm not sure if I'm taking the right steps to help just one *kachua*, never mind the rest of the environment."

"You know, I have some steps I use at work to help me solve problems," Mother said. "They might help you, too."

"What are they?" I asked.

"They're called the engineering design process. It's a series of steps that engineers use to help solve problems," she said. "You've already started by asking lots of questions about water pollution and the ways that people purify their water."

"That's true!" I said, brightening. "Father and I talked about the water cycle and pollution, and then I asked you and Dr. Gadgil about different ways to purify water."

"It sounds like you're ready for the next step— imagining different filter designs," Mother said. "Then, when you feel like you're ready, you make a plan for your filter and create it."

"But what if I test it and it doesn't work?" I asked.

Mother patted my shoulder. "Then you'll go on to the next step—improve. You can just keep improving until you've got something that is the best you can make it."

I nodded. Making a filter might take a lot of thinking and testing, but I had to try to save that turtle. I spent the rest of the day imagining possibilities.

Chapter Five | Designing a Filter

The next day, I started to gather some materials to use for my filter. I found an old shirt in my room—I thought I could try running water through that to strain out big pieces. Then I found some cheesecloth that Mother uses to make *paneer*. Just as I was about to run to the river to fill a bottle with water, I bumped into Father.

"Where are you off to?" Father asked. I told him about the filter materials that I had already gathered.

"I have an idea," Father said, "While you're at the Ganges, why don't you gather some sand from the riverbank? We could try using that to filter the water, too."

"Sand?" I asked. "Won't sand make the water dirty?"

"I think it's worth testing," Father said. "When you pack sand together there are little spaces between the grains.

It could trap some of the dirt and oil in the water and screen it out."

"Screen!" I cried. "I have some mosquito netting in my room that's a really fine screen. We can try that, too."

"Now you're thinking!" said Father. "When you get back from the river, I'll help you test your ideas, if you would like."

....

This is a very important trip, I thought to myself as I walked along. *After I fill this bottle with water and filter it, I'll be able to fill a tank for the turtle.* A few moments passed, and I couldn't ignore the question buzzing around my mind: *What about all the other animals living in the river and in other parts of the environment?*

I thought for a few moments about what Mother had said about how big the problem of pollution was. "By thinking about pollution and ways to clean up the environment," I said out loud, "I am taking little steps towards big changes. And if lots of people take little steps, it will add up to big changes—like stopping pollution from happening in the first place."

When I got to the river, I knelt down and scooped water into the bottle. As I lifted it up, sunlight streamed through the water, making it glow like amber. I remembered the microbes we had seen through the microscope in Mother's lab and wondered how many little creatures were inside the bottle.

....

When I got home, Father helped me set up a few different filters. We decided to test the mosquito netting first.

"Ready, set, pour!" I called.

I watched the water pour through the filter and settle in the bottom of the cup. "The water still looks kind of brown," I said, turning to Father.

"This is just the first step," said Father. "Remember what your mother said about the engineering design process?"

I nodded. "I guess we should test everything before I come up with a plan."

After Father and I had tested all the materials on their own,

I imagined combinations of materials. I layered sand between fabric and cheesecloth. I tried running water through pebbles and cotton. Then I drew a plan and created my filter design. It didn't work exactly the way I thought it would, so I made improvements by experimenting with more materials and combining materials. When Father and I thought we had improved our filter so that it was the best one possible, I began filling a tank with the clean water.

"You've worked very hard on this project, Salila," Father said. "I think the turtle will be happy in her new home."

"I hope so," I said. "Now I just have to find her!"

Chapter Six | Hope for the Ganges

The next day I sat on the bank of the Ganges, waiting patiently for the turtle to appear. Eventually, the *kachua* poked her head out of the water, and climbed onto the riverbank in almost the same spot where I'd seen her during the rainstorm. I carefully lifted her up and looked out at the river.

There were people using the water to wash, and boats traveling. As usual, the factory on the opposite bank blew billows of gray smoke across the sky.

But today there was also another person using the river. It was a woman, performing a ritual cleansing with the sacred water of the Ganges. Dressed in an elegant, colorful *sari*, the woman scooped water into a jug and let the water

pour out of the container, falling in an arc back into the river.

It's so beautiful, I thought. *I wish more people were here to see this.* With that simple thought, I knew how I would help not just this one turtle, but the whole Ganges.

When I got home, Mother was working in the kitchen. I placed the turtle in her new tank and smiled broadly.

"Salila," said Mother, "You did a great job. I'm proud of you."

"Thanks," I said. "I'm not done yet, though. If we

want the river to be clean, we need to stop pollution in lots of different areas, not just in the river. We need to stop polluting the air and land, too. I know I can't do it all by myself, but that's okay. I'm going to talk to my friends about everything I learned from you and Dr. Gadgil. I want to show them what I did for my new turtle. And then I'm going to ask them to help me prevent pollution."

"That's a great idea!" said Mother. "If your friends start to take care of the environment, then it will spread to their friends, and their friends will tell even more people."

"And maybe someday," I said, "if we all work really hard, I can bring the turtle back to the river so she can live there again. Together we can make a big difference in reducing pollution. The Ganges can be cleaner for the animals and people that depend on it!"

Design a Water Filter

You can design a water filter, just like Salila. Imagine that there is a pond by your house and you would like to filter some of its water. Your goal is to make the water clearer and cleaner than it was before it was filtered!

Materials

☐ Two empty soda bottles with caps
☐ Clear packing tape
☐ Clear plastic cups or a basin
☐ Loose tea
☐ Potting soil
☐ Coffee filters
☐ Cheesecloth
☐ Sand or gravel
☐ Scissors
☐ Hammer and nail, or drill

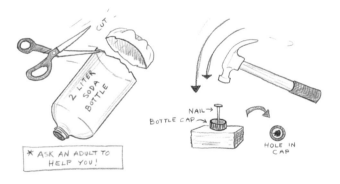

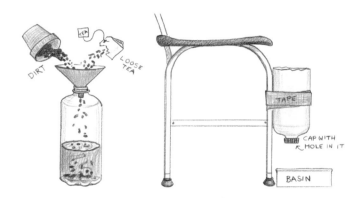

Set Up

Ask an adult to help you cut the bottom off an empty soda bottle and make a small hole in the cap with a hammer and a nail or with a drill. Put the cap back on the bottle and tape it to a chair leg. You can design your filter to fit in this bottle.

To make your polluted water, put a few tablespoons of loose tea in a whole, empty soda bottle along with hot water. Put the cap on the bottle and shake it up. Let it sit until the water is dark brown. Add a cup of potting soil and shake it up.

Test the Materials

Fill the cut soda bottle with materials that you think might help filter the pollutants out of your water. Test each material by itself. How does the water look after you run it through each material once? Twice? Does the water flow through each material at the same speed or at different speeds?

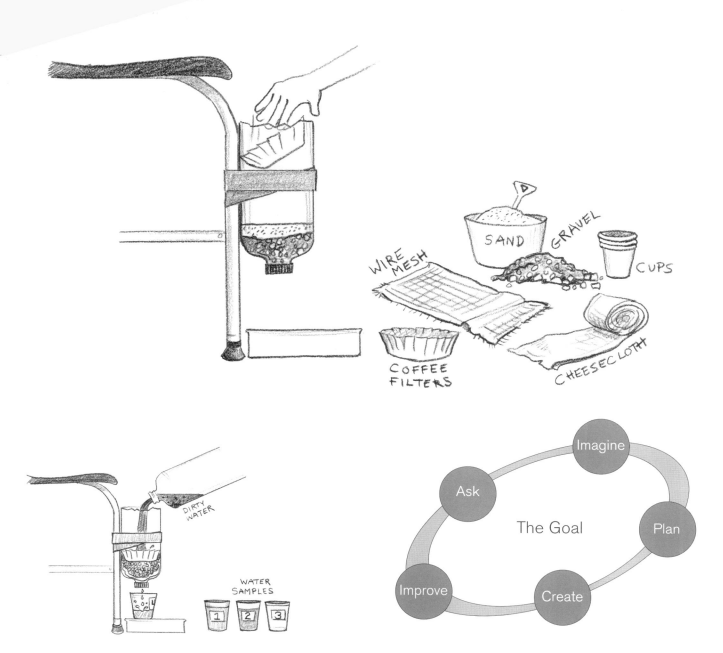

Design Your Filter

Use what you learned in your tests to make a plan for your final filter system. Think about how much of each material you will need, and how you will assemble the filter so your design is the best it can be.

Improve Your Filter

Use the engineering design process to improve your filter. Use the Internet or go to the library to learn more about different types of water filters. Can you think of more filter materials that you would like to test? Can you design a filter to help clean a different type of polluted water?

Glossary

Bacteria: Tiny, one-celled organisms that are too small to see without a microscope. Bacteria are neither animals nor plants.

Chlorine: A chemical that is used to kill microbes in water.

Contaminant: Anything found in the environment that makes it unclean or unsafe.

Engineer: A person who uses his or her creativity and understanding of mathematics and science to design things that solve problems.

Engineering design process: The steps that engineers use to design something to solve a problem.

Environment: The natural world in which we live.

Environmental engineering: The field of engineering concerned with solving problems with air, water, soil, and the natural environment.

Evaporation: The process of changing from a liquid into a gas.

Filter: Anything through which a mixture can be passed to remove one or more components.

Glacier: A huge, slow-moving body of ice.

Kachua: The Hindi word for turtle. Pronounced *ka-CHEW-ah*.

Microbes: Living things that are too small to see without a microscope. Bacteria are one type of microbe.

Monsoon: A wind that brings summer rain to South Asia.

Paneer: A type of cheese common in South Asian cuisine. Pronounced *PAN-ear*.

Pollution: Harmful substances in the soil, water, or atmosphere.

Sari: A piece of cloth that is draped over the body and worn as a garment by Hindu women. Pronounced *SAH-ree*.

Technology: Any thing, system, or process that people create and use to solve a problem.

Ultraviolet light: A type of light not visible to the human eye. It can burn living things.

Water purification: The process of removing contaminants from water.

Water vapor: Water that has evaporated and is in the form of a gas.